Grade 3

Acknowledgments
Product Development: Margaret Fetty
Editor: Michelle Howell
Cover Illustration: Tad Herr
Design and Production: Creative Pages, Inc.
Production Supervision: Sandy Batista

ISBN 1-59137-115-5
Options Publishing Inc.
P.O. Box 1749
Merrimack, NH 03054-1749
www.optionspublishing.com
Phone: 800-782-7300 Fax: 866-424-4056

Dear Parent,

Summer is a time for relaxing and having fun. It can also be a time for learning. *Summer Counts!* can help improve your child's understanding of important skills learned in the past school year while preparing him or her for the year ahead.

Summer Counts! provides grade-appropriate practice in subjects such as reading, language arts, vocabulary, and math. The ten theme-related chapters include activities and puzzles to motivate your child throughout the summer.

When working through the book, encourage your child to share his or her learning with you. You may want to tear out the answer key at the back of the book and use it to check your child's progress. With *Summer Counts!* your child will discover that learning happens anytime—even in the summer!

Apreciados padre,

El verano es una época para descansar y divertirse. También puede ser una época para aprender. *Summer Counts!* puede ayudar a que su hijo(a) mejore las destrezas importantes que aprendió el pasado año escolar al mismo tiempo que lo(a) prepara para el año que se aproxima.

Summer Counts! provee la práctica apropiada para cada grado en las asignaturas como la lectura, las artes del lenguaje y las matemáticas. Los diez capítulos temáticos incluyen actividades y rompecabezas que motivarán a su hijo(a) durante el verano.

Cuando trabaje con el libro, anime a su hijo(a) a que comparta lo que ha aprendido con Ud. Si Ud. desea puede desprender la página de las respuestas que aparece en la parte trasera del libro. Puede usar la misma para revisar el progreso de su hijo(a). ¡Con *Summer Counts!* su hijo(a) descubrirá que el aprendizaje puede ocurrir en cualquier momento—inclusive en el verano!

Table of Contents

CHAPTER 1

The Shoemaker and the Elves

Once upon a time, there was a poor shoemaker. He had only enough leather to make one pair of shoes. One evening, the shoemaker cut out the leather for the shoes. "I will finish the shoes in the morning," he said. Then he and his wife went to bed.

The next morning, the shoemaker woke to find that the shoes had been made. They were the finest shoes he had ever seen. Soon a customer came into his shop and bought the fine pair of shoes. The shoemaker now had enough money to buy leather to make two pairs of shoes. That evening, the shoemaker cut out the leather. Then he and his wife went to bed. What do you think happened?

The next morning, the shoemaker found that the shoes had been made again. These shoes were just as fine as the ones before. Soon two customers came into the shop. They each bought one pair of shoes. The shoemaker now had enough money to buy leather to make four pairs of shoes. That evening, the shoemaker cut out the leather. Then he and his wife went to bed. This happened day after day after day. The shoemaker was no longer poor.

One night the shoemaker's wife had an idea. "Let's hide in the closet and see who comes in to make these fine shoes." So they did.

At midnight, two tiny elves entered the shop from the chimney. They did not have any shoes on. The elves went right to work making shoes. They did not stop until they were done.

When the elves left, the shoemaker had an idea. "I want to thank the kind elves for helping me. I will make some shoes for them."

The next day, he sewed tiny little shoes. He left them on the table. When the elves came that night, they were surprised. They put on the little shoes and danced right up the chimney. The shoemaker did not see the elves again.

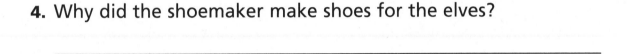

The Shoemaker and the Elves

Directions Using what you have just read, answer the questions.

1. Where did this story take place?

_____

2. What happened every time a new pair of shoes were made?

_____

3. Why do you think that the shoemaker and his wife hid in the closet?

_____

_____

4. Why did the shoemaker make shoes for the elves?

_____

5. Is this selection real or make-believe? How do you know?

_____

_____

Nouns: Naming Words

REMEMBER

Naming words are called **nouns**. Nouns name people, places, and things.

Singular nouns name one person, place, or thing. **Plural nouns** name more than one person, place, or thing. Some nouns change in spelling when they become plural.

- Add **-s** to most singular nouns to make them plural.
- Add **-es** to singular nouns that end with **s**, **x**, **ch**, and **sh**.
- Change **y** to **i** and add **-es** to nouns that end with a consonant and **y**.

EXAMPLES

Singular	Plural
one girl	two girls
one fox	two foxes
one family	two families
one child	two children

Helping Out at the Farm

Directions Read the sentences and the nouns that follow. Write the correct form of each noun on the line.

1. Two _____ visited a farm to help out the farmer. (class)

2. The students saw many animals in the _____. (barn)

3. They wanted to help feed the _____. (animal)

4. Anna got to feed a _____. (horse)

5. Ernie got to feed four _____. (puppy)

6. All the students got to feed the _____. (chicken)

Neighborhood Helpers

Directions Write the singular form of each plural word.

1. nurses   _____
2. policemen   _____
3. teachers   _____
4. coaches   _____
5. firemen   _____

Grandmother's School

Directions Read the story. Use words from the box below to fill in the blanks.

art	study	students
classes	teacher	room

My grandmother went to school in a one-room schoolhouse. All the students sat together for their **(6)** _____. The older students helped the younger ones with their work. At the same time, the **(7)** _____ would show the older students how to do a new lesson. All the children would **(8)** _____ together in the same **(9)** _____. Grandmother said that sometimes it was very noisy in the classroom, but she loved it. Her favorite classes were **(10)** _____ and math.

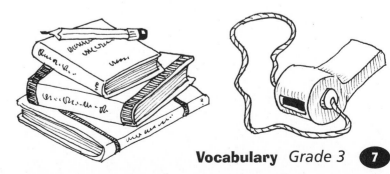

Counting Whole Numbers

REMEMBER

We count **whole numbers**. Whole numbers can be written as **digits** or **words**. There are 10 digits in our number system: 0, 1, 2, 3, 4, 5, 6, 7, 8, and 9.

EXAMPLES

digits	words
0	zero
12	twelve
100	one hundred
1,000	one thousand

Whole Number Fun

Directions Write out each number in words.

1. 4 _____

2. 8 _____

3. 15 _____

4. 11 _____

5. 14 _____

6. 70 _____

7. 200 _____

8. 3,000 _____

9. 30 _____

10. 60 _____

Directions Write the missing numbers on the lines.

11. 1, 2, _____, 4, _____, 6, 7, _____, _____, 10

12. 2, 4, _____, 8, _____, 12 _____, 16, _____, _____

13. 3, 6, 9, _____, 15, _____, _____, 24, 27, _____

14. 10, _____, 30, _____, 50, _____, 70, _____, _____, 100

15. 5, _____, _____, 20, 25, _____, 35, 40, _____, _____

Helping the Earth

Directions The students at Jefferson Elementary School learned how recycling helps the earth. They started recycling paper at their school. The graph below shows the weight of the paper each class collected. Use the graph to answer the questions.

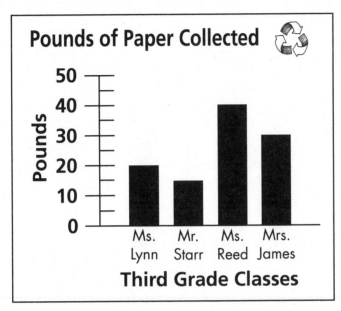

1. Whose class collected the most paper? _____

2. Whose class collected the least paper? _____

3. Whose class collected about 30 pounds of paper? _____

4. How many pounds of paper did Mr. Starr's class collect? _____

5. About how many pounds of paper did all four classes collect? _____

6. Ms. Reed's class wants to collect a total of 50 pounds of paper. About how much more paper will the class need to collect? _____

Helper Word Search

Directions Find the words listed in the box. Then circle them in the puzzle. The words can be hidden across, down, or diagonally.

brother	teacher	mother	student
sister	friend	father	neighbor

```
s   l   m   o   t   h   e   r
t   e   a   c   h   e   r   f
u   d   s   c   k   n   d   a
d   f   r   i   e   n   d   t
e   t   h   g   s   a   g   h
n   s   e   c   a   t   w   e
t   b   r   o   t   h   e   r
n   e   i   g   h   b   o   r
```

Thanks for the Help!

Direction Think about a person who has helped you. Who was it? What did this person do? How did it make you feel? Write a letter to that person thanking them for their help.

Dear _____,

_____

_____

_____

_____

_____

_____

_____

_____

_____

_____

_____

_____

_____

_____

_____

From: _____

CHAPTER 2

What a Day for a Knight!

"Did you hear that noise?" Jenna asked her twin brother Kyle. CREAKKKK! This time the noise got louder.

"Yeah, I heard it, too," Kyle said. "What do you think it is?"

"It came from behind that large boulder up on the hill," Jenna said. She pointed to a large rock about thirty feet away. "Let's go find out what it is."

Kyle's eyes widened. "I don't think it is a good idea. Mom said not to go too far from the campground."

"We won't go too far. I want to see what made that noise," said Jenna.

Kyle followed his sister. As they climbed up the hill, the noise got louder. When they got to the top, they both gasped. Below them was a beautiful white horse. Next to the horse was a knight in armor. The knight was having trouble walking. Each time he moved, his armor made a creaking sound.

"Can't anybody help me?" the knight shouted toward the sky.

Kyle and Jenna couldn't believe their eyes. "We can!" they both replied.

The knight looked at them. Then a man appeared, shouting, "Cut! Stop the film."

"What are you kids doing here?" the man asked.

"We were hiking in the woods when we heard a strange noise," Jenna said. "We climbed up here to see what it was. We never expected to find a knight!"

The man laughed. "This isn't a real knight. We're filming a movie. Would you like to stay and watch?"

"Wow! Our friends will never believe we got to see a movie being filmed while on a camping trip in the woods!" said Kyle.

What a Day for a Knight!

Directions Using what you have just read, answer the questions.

1. Where does this selection take place?

   _____

2. What made the strange noise that Kyle and Jenna heard?

   _____

3. What did the man filming the movie do after he heard Jenna and Kyle offer to help?

   _____

   _____

4. Which of the following sentences is an opinion? Circle it.

   The knight was having trouble walking.

   Below them was a beautiful white horse.

5. What would be another title for this selection?

   _____

   _____

Nouns: Common and Proper

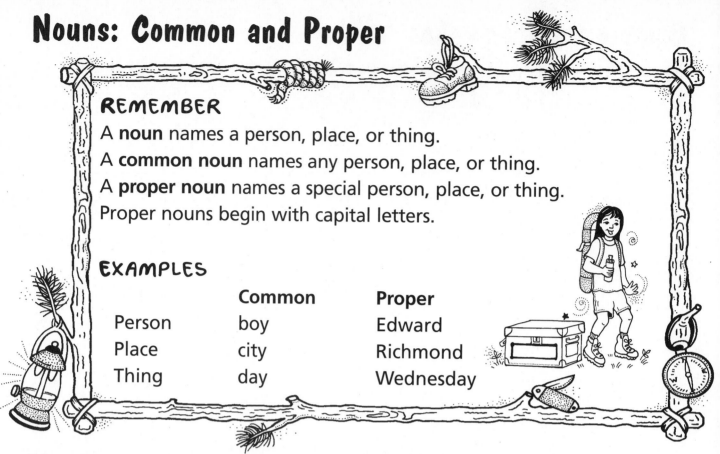

REMEMBER

A **noun** names a person, place, or thing.
A **common noun** names any person, place, or thing.
A **proper noun** names a special person, place, or thing.
Proper nouns begin with capital letters.

EXAMPLES

	Common	Proper
Person	boy	Edward
Place	city	Richmond
Thing	day	Wednesday

Directions Write the common nouns and the proper nouns in each sentence. Remember to begin each proper noun with a capital letter.

	Common	Proper
1. Rita is my friend.	_____	_____
2. She left for camp on sunday.	_____	_____
3. The camp is in missouri.	_____	_____
4. It is near the town of branson.	_____	_____
5. She will stay at camp kano for a month.	_____	_____
6. Her parents will pick her up on the fourth of july.	_____	_____

Camping Comparisons

Directions Analogies are made from pairs of words that have the same relationship. Complete the analogies using words from the box below.

vacation	camp	tent	boots	beach
family	trail	fishing	forest	lake

1. HILL is to MOUNTAIN as POND is to _____.

2. SKI is to SKIING as FISH is to _____.

3. FLOWERS are to GARDEN as TREES are to _____.

4. HAT is to CAT as REACH is to _____.

5. DRIVE is to ROAD as HIKE is to _____.

Going on Vacation

Directions Read the letter. Use words from the box above to fill in the blanks.

Dear Kim,

I am so excited. My **(6)** _____ and I are going to

(7) _____ at a park. Mom bought a new

(8) _____ for us to sleep in. She also bought good

(9) _____ for our feet so we can go hiking. Dad said we

would take our canoe out on the **(10)** _____. I think I will

take my rod and reel with me in case we go **(11)** _____.

This will be the best **(12)** _____ ever!

Your friend,

Pam

Place Values

REMEMBER
Place value tells the value of a digit. The value of the digit depends on its place in the number.

EXAMPLE

1,465 has four digits.

1 is in the **thousands** place   **thousands** = 1,000s
4 is in the **hundreds** place   **hundreds** = 100s
6 is in the **tens** place   **tens** = 10s
5 is in the **ones** place   **ones** = 1s

Take Your Places

Directions Write the value of each underlined digit.

1. 4<u>7</u> _____

2. 5<u>2</u>6 _____

3. <u>4</u>,197 _____

4. 5<u>3</u>8 _____

5. <u>2</u>,473 _____

Directions Write the number.

6. four hundreds, two tens, and six ones _____

7. eight hundreds, five tens, and two ones _____

8. one thousand, eight hundreds, one ten, and nine ones _____

9. two thousands, six hundreds, and four ones _____

Name the Numbers

Directions Use the numbered tents below to answer the questions. Use each number only once for each question.

1. What is the smallest 2-digit number that can be made with the tents? _____

2. What is the largest 2-digit number that can be made with the tents? _____

3. What is the smallest 3-digit number that can be made with the tents? _____

4. What is the largest 3-digit number that can be made with the tents? _____

5. Using all three tents, write an even number. _____

6. Using all three tents, write an odd number. _____

7. Can you make more even or odd numbers with the tents? Why?

_____

_____

_____

Camping Puzzle

Directions Write words from the box below to complete the puzzle.

fire   lake   boots   trail
hike   tent   woods   backpack

ACROSS

3. A bag used when hiking

5. A path through the woods

6. A place with trees

7. A cloth shelter for camping

DOWN

1. A long walk through the forest

2. A body of water

3. Sturdy shoes used for hiking

4. You light this to stay warm

Pack Your Bags

Suppose you were going camping. What activities would you do? How would you pack? Write a list of things you would pack in your bag for the trip.

CHAPTER 3

The Cricket and the Ant

Cricket sat in the grass singing. He was enjoying the warm summer day when Ant marched past. Ant was carrying a piece of grain on his back.

"Hello, Ant," called Cricket. "It's a beautiful day, isn't it? Will you stop and sit with me?"

"No time to stop," Ant said. "I have too much work to do. I have to store this grain so that I will have food for the winter."

"Silly Ant!" Cricket laughed. "Winter is months away. Sit with me now, and I will help you later."

"No thanks," Ant replied. "The more grain I store, the more food I will have for winter."

Cricket shook his head as Ant continued on his way. "Well, I am not going to waste this perfect day," he said. He started to sing again. In fact, Cricket sat in the grass and sang all summer long.

Soon, autumn came and the weather turned cooler. Still Cricket sang. "I have plenty of time," he thought to himself.

One day, the weather got especially cold. Snow began falling. Winter had arrived. "Oh, no," Cricket gasped. "I have waited too long!" Cricket hopped around the ground looking for food. There was nothing to be found. He rushed to Ant's house. "Please help me, Ant. I am hungry," Cricket begged.

"Ha! Who is singing now?" asked Ant. "I have just enough food for me. I am sorry, but I cannot help you."

It was a hard winter for Cricket. He learned his lesson. The following summer, he worked even harder than Ant collecting food for the winter.

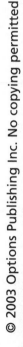

The Cricket and the Ant

Directions Using what you have just read, answer the questions.

1. In this selection, how is Cricket different from Ant?

_____

_____

2. Why does Ant work in the summer?

_____

_____

3. What happens after the snow begins to fall?

_____

4. What lesson do you think Cricket learned? Explain your answer.

_____

_____

_____

5. Is this selection real or make-believe. How do you know?

_____

_____

Verb Tenses

REMEMBER

A **verb** shows action. It also shows time. The tense of a verb tells you when something happened.

A verb in the **present tense** shows action that is happening now. Many of these verbs tell about one person, place, or thing and end in **-s**.

A verb in the **past tense** shows action that has already happened. Many of these verbs end in **-ed**.

EXAMPLES

Present Tense

Seals often rest together in groups.

Past Tense

The seals rested in the sun yesterday.

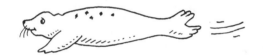

Directions Draw a line under the verb in each sentence. Write if it is present or past tense on the lines.

1. The sea lion paddled with its flippers. _____

2. Sea lions walk on all four flippers. _____

3. The sea lion pup follows its mother. _____

4. An elephant seal flipped sand on its back. _____

5. Harp seals open their eyes under water. _____

6. The seals swim through the water. _____

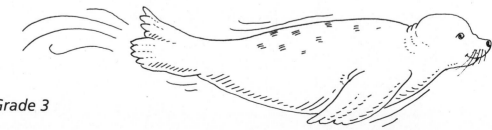

Animal Words

Directions Cross out the word in each group that does not belong.

1. seal   dolphin   cat

2. bark   swim   roar

3. kittens   horse   geese

4. pond   duck   elephant

Pond Life

Directions Read the story. Use words from the box below to fill in the blanks.

pond	ducks	geese	bread

In the winter, my brother and I ice skate on a small

(5) _____ near our house. In the spring and summer, it is

full of **(6)** _____ and **(7)** _____ that swim

and splash around. We sit under a tree and watch them play. Sometimes

we feed them **(8)** _____ .

Dollars and Cents

REMEMBER

Dollars and **cents** are written with special symbols.

EXAMPLE

dollar sign → ← decimal point

$1.75

dollars → ← cents

one dollar and seventy-five cents

Cents can be written using only the cent sign or ¢.

15 cents is written 15¢.

Directions Write the value of the money.

1.

2 quarters, 1 dime,
1 nickel, 3 pennies _____

2.

3 quarters, 2 dimes,
1 nickel _____

3.

6 one dollar bills,
1 quarter, 1 penny _____

4.

3 one dollar bills,
5 dimes, 1 nickel,
2 pennies _____

Coin Count

Directions Mei Li has several coins in her purse. She will not tell what coins she has. She gives the following clues: (1) She has a total of 47¢. (2) Her largest coin is a quarter. Use this to answer the questions.

1. Do you think Mei Li has any pennies? Why or why not?

   _____

   _____

2. Mei Li gives another clue. She says that she has only one quarter. What is the largest number of coins she could have? Name them.

   _____

   _____

   _____

3. What is the smallest number of coins she could have? Name them.

   _____

   _____

   _____

4. Mei Li gives another clue. She says that she has 6 coins. What are they? Tell how you know.

   _____

   _____

   _____

Animal Code

Directions Look at the codes. Each number stands for a letter. Write the letters on the blanks to answer the riddles.

A 1	B 2	C 3	D 4	E 5	F 6	G 7	H 8	I 9	J 10
K 11	L 12	M 13	N 14	O 15	P 16	Q 17	R 18	S 19	T 20
U 21	V 22	W 23	X 24	Y 25	Z 26				

1. What did the duck say when he could not pay?

‾‾‾ ‾‾‾ ‾‾‾   ‾‾‾ ‾‾‾   ‾‾‾ ‾‾‾   ‾‾‾ ‾‾‾
16 21 20   9 20   15 14   13 25

‾‾‾ ‾‾‾ ‾‾‾ ‾‾‾ .
2   9   12 12

2. Why did the baby birds fly home when their mother came back with the groceries?

‾‾‾ ‾‾‾ ‾‾‾   ‾‾‾ ‾‾‾ ‾‾‾   ‾‾‾ ‾‾‾ ‾‾‾ ‾‾‾ ‾‾‾ ‾‾‾ .
19 8   5   8   1   4   20 23 5   5   20 19

3. How did the farmer know that the horse got loose?

‾‾‾ ‾‾‾ ‾‾‾   ‾‾‾ ‾‾‾ ‾‾‾   ‾‾‾ ‾‾‾ ‾‾‾ ‾‾‾ ‾‾‾ ‾‾‾ ‾‾‾ ‾‾‾ .
20 8   5   16 9   7   19 17 21 5   1   12 5   4

4. What did the bird say at the big sale?

‾‾‾ ‾‾‾ ‾‾‾ ‾‾‾ ‾‾‾ ,   ‾‾‾ ‾‾‾ ‾‾‾ ‾‾‾ ‾‾‾ !
3   8   5   1   16   3   8   5   1   16

Drawing Contest

Imagine your favorite wildlife magazine is having a drawing contest. The winning drawing will be used as the next cover. The directions say to draw a picture of your favorite wildlife scene. Design your magazine cover.

CHAPTER 4

Twice a Hero

On Saturday afternoon, Nick Sanchez became a hero—not once, but twice. Nick had just finished playing ball with some friends at Handell Park. As he left the park, he noticed a small boy playing in the street. There were no other adults around. A truck suddenly appeared in the street, moving toward the boy. The boy did not see the truck coming.

Nick jumped into action. He ran into the street, grabbed the boy's hand, and pulled him out of the path of danger. Nick's quick action saved the boy. The boy's parents called Nick a hero. But this is not the end of the story.

On his way home, Nick turned the corner and spotted thick, black smoke. It was coming out of the window of an apartment building. Nick knew something was wrong. He quickly found a fire alarm and pulled the handle. The fire trucks rushed in and put out the fire. No one was hurt. Everyone cheered for Nick. Nick was a hero for the second time that day!

When asked what he was going to do next, Nick replied, "I'm going to try to get home before anything else happens!"

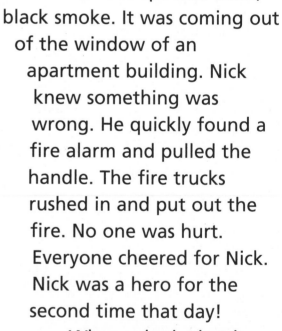

Twice a Hero

Directions Using what you have just read, answer the questions.

1. Number the following sentences **1**, **2**, **3**, or **4** in the order that they happened.

    _____ The boy's parents called Nick a hero.

    _____ Nick found a fire alarm and pulled the handle.

    _____ Nick noticed a small boy playing in the street.

    _____ Fire trucks appeared on the scene.

2. Who cheered for Nick?

    _____

3. What do the words **path of danger** mean?

    _____

    _____

4. How did Nick know something was wrong at the apartment?

    _____

    _____

5. Do you know someone who can be called a hero? Write a sentence or two about that person below.

    _____

    _____

    _____

Verbs: Past and Present

REMEMBER

Many verbs show action, but some do not. One example is the verb **be.** It tells what something is or is like.
Am, **is**, and **are** tell about things that are happening now.
Was and **were** tell about things that happened in the past.

EXAMPLES

Happening Now
I **am** ready.
She **is** ready.
They **are** ready.

Happened in the Past
I **was** ready.
She **was** ready.
They **were** ready.

Read All About It

Directions Circle the correct verb in each sentence.

1. The *Trumpet* (is,   are) a small newspaper.

2. It (was,   were) my idea.

3. Being a newspaper editor (is,   are) a big job.

4. There (is,   are) a million things to do.

5. The first weeks (was,   were) really hard.

6. I (was,   were) the only reporter.

7. We (is,   are) proud of our neighborhood newspaper.

Word News

Directions Complete the analogies using words from the box below.

reporter	hello	word	message	newspaper
sent	news	letter	write	reading

1. LEAVE is to GOODBYE as GREET is to _____.

2. TELEVISION is to WATCHING as BOOK is to _____.

3. GAVE is to TOOK as RECEIVED is to _____.

4. ALPHABET is to LETTER as SENTENCE is to _____.

Here's the Latest

Directions Read the story. Use words from the box above to fill in the blanks.

There are many ways that people get information about what is happening in the world. They can watch the **(5)** _____ on television to get the latest information. They can get a

(6) _____ and read the articles. They can get on a computer to get an e-mail **(7)** _____. Some people like to sit down with a pen and paper and **(8)** _____ a

(9) _____ to a friend. They act as their own

(10) _____!

Comparing Numbers

REMEMBER

When you compare numbers, you find which number is greater than or less than. Look at the digits in the highest place value to compare. Use the signs > and < .
The sign > means **greater than**, or larger.
The sign < means **less than**, or smaller.

EXAMPLES

Which is greater than?

41   35
Compare 4 and 3.

41 > 35 because 4 > 3.

Which is less than?

129 154
Each begins with 1,
so compare 2 and 5.

129 < 154 because 2 < 5.

Directions Write < or > between each pair of numbers.

1. 37 _____ 41

2. 92 _____ 88

3. 125 _____ 148

4. 419 _____ 391

5. 1,218 _____ 1,243

6. 3,256 _____ 3,380

Directions Look at the street map. Only two addresses are shown. Using the Address List, write the address on each of the other ten buildings. Hint: Addresses get larger as you move away from downtown.

7.

					716

Walnut Avenue

987					

Odd numbers go on this side of Walnut Street

Downtown

Address List

724	943
820	882
984	835
879	777
709	926

The Temperature Is Rising

Directions The chart below shows the daily temperatures for Chicago as reported on the Internet. Use the chart below to answer the questions.

Temperatures on July 16th

Time	Temperatures
10:00 A.M.	68° F
11:00 A.M.	72° F
12 Noon	76° F
1:00 P.M.	
2:00 P.M.	

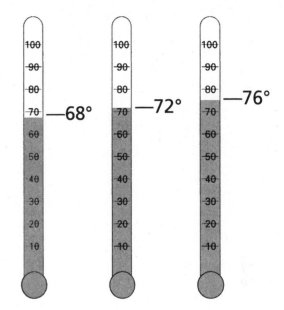

1. If the temperature continues to rise at the same rate, what will the temperature be at 1:00 P.M.? _____

2. What will the temperature be at 2:00 P.M.? _____

3. During the day, will the temperature reach 90°F? Why or why not?

   _____

   _____

4. How high will the temperature most likely reach? Explain your answer.

   _____

   _____

   _____

Writer's Scramble

Directions Read each clue. Unscramble the letters to find the answers. Write the words on the lines. Then use the numbered letters to solve the riddle.

1. A short letter

eton $\underline{}_{1}$ $\underline{}$ $\underline{}$ $\underline{}$

2. Something sent in the mail

telret $\underline{}$ $\underline{}_{2}$ $\underline{}$ $\underline{}$ $\underline{}$ $\underline{}$

3. To put words on paper

twire $\underline{}$ $\underline{}$ $\underline{}_{3}$ $\underline{}$ $\underline{}$

4. An e-mail letter

smagsee $\underline{}$ $\underline{}$ $\underline{}$ $\underline{}_{4}$ $\underline{}$ $\underline{}_{6}$ $\underline{}$

5. Something to write with

clepin $\underline{}_{5}$ $\underline{}$ $\underline{}$ $\underline{}$ $\underline{}$

6. To look at words and understand them

drea $\underline{}_{7}$ $\underline{}$ $\underline{}$ $\underline{}$

Now solve the riddle: What is black and white and read all over?

$\underline{}_{6}$ $\underline{}_{1}$ $\underline{}_{2}$ $\underline{}_{3}$ $\underline{}_{4}$ $\underline{}_{5}$ $\underline{}_{6}$ $\underline{}_{5}$ $\underline{}_{2}$ $\underline{}_{7}$

Read All About It!

Directions Imagine you are a newspaper reporter. Think of an exciting event that has happened recently. Write a story about it for a newspaper. Be sure to answer the questions who, what, when, where, why, and how in your story.

THE TRUMPETER

_____

_____

_____

_____

_____

_____

_____

_____

_____

_____

_____

_____

_____

_____

_____

_____

CHAPTER 5

Special Doctors

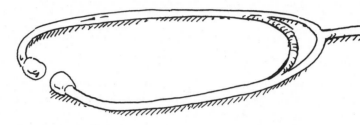

Did you know that some doctors only take care of children? That is because children have different needs than adults.

Children grow tall and strong very quickly. They must eat the right foods to grow up healthy. Adults do not need the same foods as children. They do not need as much of it, either. Children need to get more calcium than adults do. Calcium makes their bones strong. You can get calcium from milk, yogurt, and ice cream.

Children have special needs when they get sick, too. They need to be treated differently than adults. Children cannot always take the same kinds of medicine. They must take smaller amounts of the medicine because their bodies are smaller. There are many differences in caring for adults than children. You can see why there are special doctors just for children!

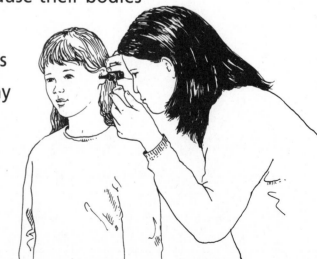

Special Doctors

Directions Using what you have just read, answer the questions.

1. What is the article mostly about?

_____

_____

2. Name two ways the article says children are different from adults.

_____

_____

3. What does calcium do?

_____

4. What is a good reason to eat ice cream?

_____

_____

5. Is this selection real or make-believe? Tell how you know.

_____

_____

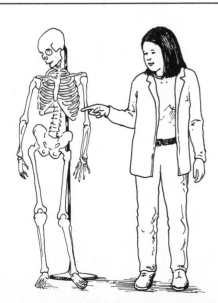

Pronouns

REMEMBER

A **pronoun** takes the place of one or more nouns.
The pronouns **I, you, he, she, it, we,** and **they** can take the place of the subject in a sentence. These are **subject pronouns**.
The pronouns **me, you, him, her, it, us,** and **them** can take the place of nouns that follow action verbs and words such as to, for, at, of, and with. These are **object pronouns**.

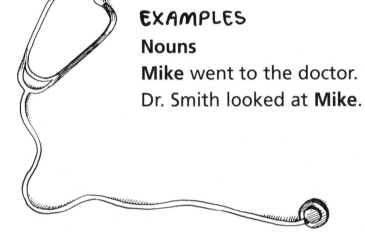

EXAMPLES

Nouns	**Pronouns**
Mike went to the doctor.	**He** went to the doctor.
Dr. Smith looked at **Mike**.	Dr. Smith looked at **him**.

Directions Write the subject or object pronoun that takes the place of the underlined word or words in each sentence.

1. The nurse is named Ms. Jacobs.   _____

2. Ms. Jacobs will give Mike a shot.   _____

3. The shot will make Mike feel better.   _____

4. Dr. Smith lives next door to Mike.   _____

5. Dr. Smith is kind to her patients.   _____

6. My sister and I see Dr. Smith when we are sick.   _____

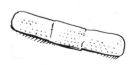

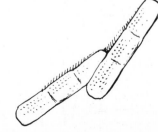

Body Parts

Directions Write the singular form of each plural word.

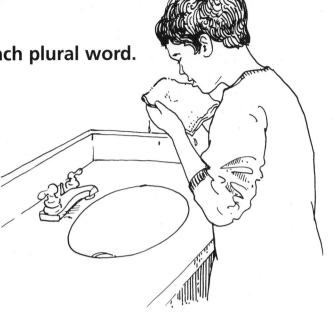

1. bones   _____

2. noses   _____

3. mouths   _____

4. ears   _____

5. eyes   _____

Staying Healthy

Directions Read the story. Use words from the box below to fill in the blanks.

teeth	calcium	sick	hands	hair	healthy

It is very important to take good care of yourself and stay

(6) _____. To keep from getting sick, you can wash

your **(7)** _____ with soap and water. You should also

wash your **(8)** _____ with shampoo. Don't forget to brush

your **(9)** _____ with a toothbrush twice a day. Make sure

you get enough **(10)** _____ to keep your bones strong.

When you are **(11)** _____, you might want to go see

a doctor.

Rounding Numbers

REMEMBER

A **rounded number** is close to an exact amount. It usually ends in a zero. Look at the number you are rounding to.

If the number to the right of that number ends in 4 or less, round down. If that number ends in 5 or higher, round up.

EXAMPLES

Number	Rounded	Reason
2,893: round to the nearest 10.	2,890	3 is less than 4, so round down.
2,893: round to the nearest 100.	2,900	9 is greater than 5, so round up.
2,893: round to the nearest 1,000.	3,000	8 is greater than 5, so round up.

Number Round Up

Directions Round each number to the nearest tens, hundreds, or thousands. Circle the correct answer.

To the nearest 10

1. 48: 40 or 50

4. 216: 210 or 220

To the nearest 100

2. 272: 200 or 300

5. 687: 600 or 700

To the nearest 1,000

3. 3,500: 3,000 or 4,000

6. 6,910: 6,000 or 7,000

Directions Round each number to the place value shown.

7. 32 years old (**nearest 10s**) _____

8. 157 pounds (**nearest 100s**) _____

9. 2,846 nurses (**nearest 100s**) _____

10. 3,501 thermometers (**nearest 1,000s**) _____

Pencil Prize

Directions Dr. Mendoza lets children pick a pencil from a bag when they leave his office. The bag has 4 red pencils and 8 blue pencils. Use this to answer the questions.

1. How many pencils are in the bag?

   _____

2. What color pencil is most likely to be picked? Why?

   _____

   _____

3. What color pencil is least likely to be picked? Why?

   _____

   _____

4. Dr. Mendoza wants the children to pick a red pencil more often than a blue pencil. What changes should he make? Explain your answer.

   _____

   _____

5. Dr. Mendoza wants it to be **twice** as likely that a child will pick a red pencil than to be **twice** as likely as picking a blue pencil. What changes should he make? Explain your answer.

   _____

   _____

   _____

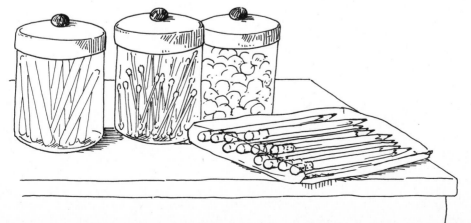

Doctor Word Search

Directions Find the words listed in the box. Then circle them in the puzzle. The words can be hidden across, down, or diagonally.

nose	ears	eyes	exercise
nurse	teeth	doctors	medicine

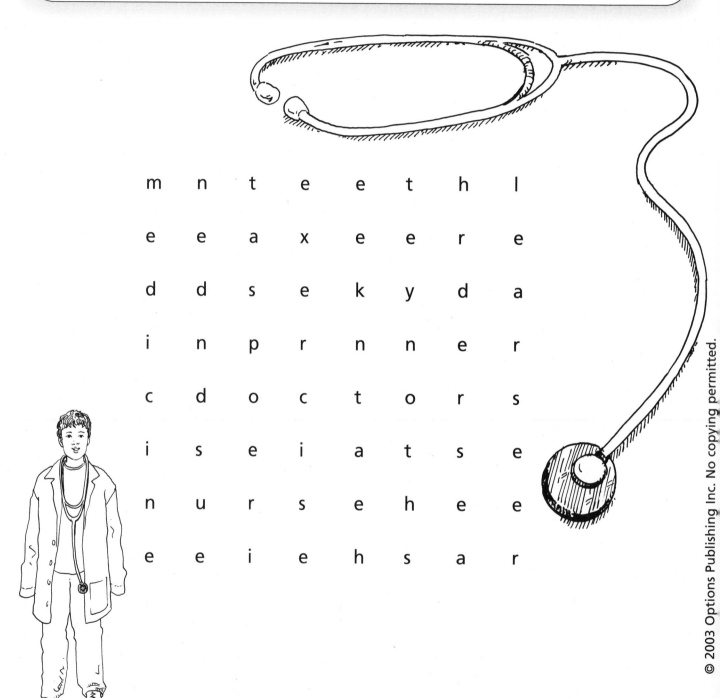

```
m  n  t  e  e  t  h  l
e  e  a  x  e  e  r  e
d  d  s  e  k  y  d  a
i  n  p  r  n  n  e  r
c  d  o  c  t  o  r  s
i  s  e  i  a  t  s  e
n  u  r  s  e  h  e  e
e  e  i  e  h  s  a  r
```

A Healthy Poem

Directions Staying healthy is important. Exercising, eating right, and getting lots of sleep help us stay healthy. Think about other ways to stay healthy. Write a two-line poem about it. Then draw a picture to go along with your poem. Here's an example:

My friends and I go out to play.
That's how we exercise every day.

CHAPTER 6

Megan's Challenge

Megan walked to the end of the diving board. Cold drops of water fell from her long hair and rolled down her back. She had been in the swimming pool all day, but had stayed away from the diving board.

"What are you waiting for, Mega-kid?" her dad called from the pool. "I'll be right here when you jump in."

Megan took a deep breath and wrapped her arms around her stomach tightly.

"Do you want me to show you how to do it?" her dad asked. He looked up at her from the water. Megan looked scared. She shook her head.

Swimming was Megan's favorite sport. She liked how the water felt. She was never afraid of the water. She jumped right in the first time her parents took her to the pool. Her dad had to jump in to help her. After that it only took a few lessons before she was swimming like a fish.

So why was this scaring her? She looked down at her dad. He believed that she could do it! Even if she was scared, she would jump. She would not quit. She would learn how to dive! Megan took a big breath and held her nose.

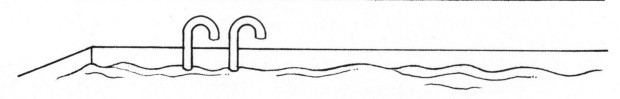

Megan's Challenge

Using what you have just read, answer the questions.

1. What is this selection mostly about?

_____

_____

2. Where does this selection take place?

_____

_____

3. Why do you think Megan had been staying away from the diving board all day?

_____

_____

4. How can you tell that Megan is scared?

_____

_____

5. Do you think Megan will overcome her fear? Why or why not?

_____

_____

Adjectives

REMEMBER

An **adjective** is a word that describes a noun. Adjectives can tell what kind or how many. They can also compare things.

Add **-er** to most adjectives to compare two persons, places, or things.

Add **-est** to compare three or more persons, places, or things. With longer adjectives, use **more** and **most** to compare a person, place, or thing.

EXAMPLES

Leah runs **fast.**

Leah runs **faster** than Thomas.

Leah is the **fastest** runner in the class.

Ling is a **skillful** pitcher.

Ling is a **more skillful** pitcher than Jess.

Ling is the **most skillful** pitcher on the team.

Directions Write the correct form of the adjective to complete each sentence.

1. Our school Fun Run is the _____ event of the year. (exciting)

2. I ran _____ this year than last. (fast)

3. The weather was _____ last year than this year. (warm)

4. The running course was the _____ that I have ever seen. (difficult)

5. My sister was the _____ runner in the race. (young)

6. I was _____ than Fred this year. (slow)

7. Fred is a _____ runner than I am. (skillful)

Winning Teams

Directions Cross out the word in each group that does not belong.

1. pass   hit   dive

2. running   jogging   sleeping

3. shoot   dribble   swing

4. bat   team   group

The Basketball Team

Directions Read the sports story. Use words from the box below to fill in the blanks.

win	run	team	catch
won	pass	sport	shoot

This basketball **(5)** _____ is the best in the city. The

players have **(6)** _____ every game this season. Watch

them **(7)** _____ this game, too! Carol Green is one of their

best players. She can **(8)** _____, **(9)** _____,

and **(10)** _____ the ball. You should see her

(11) _____ up and down the court. Carol plays with her

teammates, not against them. Together, they make this

(12) _____ very exciting.

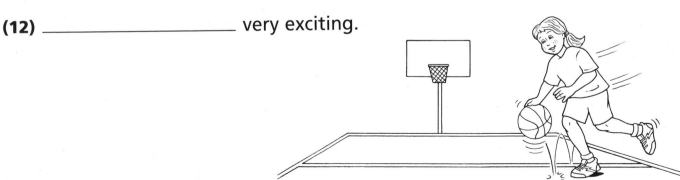

Adding Whole Numbers

REMEMBER

When you **add**, you put groups of things together. To show addition, you write +. This is called a **plus sign** or an **addition sign**. The answer is called the **sum** or **total**. Sometimes, you will need to **regroup**, or **carry**, to another place value.

EXAMPLES

```
  1
 26
+18
───
 44
```

```
 $8.83
+$4.05
──────
$12.88
```

```
 11      ←── regrouped numbers
259
+148
────
407
```

Find the Sum

Directions Find each sum. Regroup if you need to.

1.
```
  58
+ 19
```

2.
```
  91
+ 27
```

3.
```
  79
+ 42
```

4.
```
 164
+  25
```

5.
```
 521
+ 317
```

6.
```
 801
+ 199
```

7.
```
 $5.78
+$2.19
```

8.
```
 $8.19
+$5.78
```

9.
```
 $7.75
+$2.19
```

Charting the Miles

Directions Juan uses a chart to keep track of the number of miles he runs each month. Use the chart below to answer the questions.

Miles Juan Runs

Month	Miles
January	19
February	21
March	18
April	22
May	
June	

1. What is the average number of miles Juan ran from January to April? Show your work.

   _____

   _____

   _____

2. What is the range in the number of miles Juan ran from January to April? Show your work.

   _____

   _____

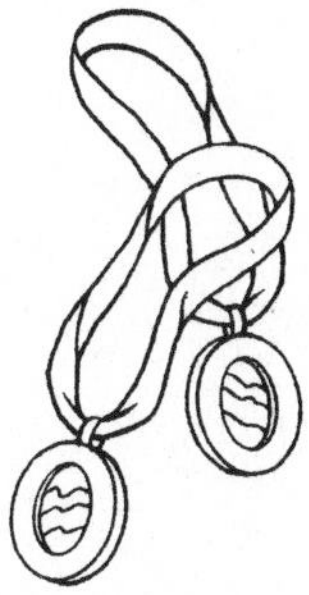

3. Estimate the number of miles Juan will run in May and June. Write the numbers on the chart.

4. What did you assume to be true when you estimated for question 3? Explain.

   _____

   _____

   _____

5. What is the total distance you think Juan will run during the whole year? Explain your answer.

   _____

   _____

   _____

Sports Riddles

Directions Use the clues to solve the riddles. Hint: Think about the words in dark print to help you.

EXAMPLE: What do you call a baseball **mitt** that is the right size?

f i t   m i t t

1. What do you call a ball that goes **high** into the air when it is hit?

h i g h ___ ___ ___ ___

2. What do you call a race that is **fun**?

f u n ___ ___ ___

3. What do you call roller skates that are a good **deal** because they are on sale?

___ ___ ___ ___ ___ d e a l

4. What do you call trying to tag a baseball player who is stealing a **base**?

b a s e ___ ___ ___ ___ ___

5. What do you call the best **skate**?

___ ___ ___ ___ ___ s k a t e

My Favorite Sport

Directions Do you have a favorite sport? What equipment do you need to play this sport? Do you wear a helmet? Do you need skates? Write a paragraph telling about your favorite sport.

CHAPTER 7

Two Twisters Touch Down!

FORT WORTH, TEXAS—On March 28, 2002, two tornadoes hit Fort Worth, Texas. The next day, crews were still at work cleaning up the city.

The tornadoes, or twisters, hit Fort Worth at 6:30 P.M. on March 28, 2002. Twisters are strong storms. They bring heavy rains and giant hail. With this storm, the strong winds knocked down power lines. Crews worked all night to get the power back on. The cleanup cost more than $450 million.

The day after the storm, Fort Worth was closed. Offices and stores were closed. The twisters damaged 50 businesses. Many buildings had to be torn down.

About 100 people lost their homes. Another 1,000 homes were damaged. Most people needed new homes.

They also needed new furniture, clothes, and other items.

Fort Worth

Two Twisters Touch Down!

Directions Using what you have just read, answer the questions.

1. What is another name for a tornado?

   _____

2. What happens during a tornado?

   _____

   _____

3. In this article, how did the twister affect the area? Name two ways.

   _____

   _____

4. How much did it cost to clean up Fort Worth?

   _____

   _____

5. Why did it cost so much to clean up after the storm?

   _____

   _____

Adjectives: Articles

REMEMBER

The special adjectives **a**, **an**, and **the** are articles. The word **a** is used before words that begin with consonant sounds. The word **an** is used before words that begin with vowel sounds. The word **the** is used before a noun that names a particular person, place, or thing.

EXAMPLES

We read **a** book about storms.
She used **an** encyclopedia for her report.
The sky was filled with dark clouds.

Directions Choose the best article to complete each sentence.

1. _____ tornado is a powerful storm. (A, An)

2. _____ winds from tornadoes are very strong. (A, The)

3. In our town, _____ alarm warns of tornadoes. (a, an)

4. _____ lights went out during the storm. (A, The)

5. _____ clap of thunder made me jump. (A, An)

6. _____ eye of the hurricane is the center of the storm. (A, The)

7. Our class read _____ article about weather. (a, an)

8. We met _____ author who wrote the book about storms. (an, the)

Weather Analogies

Directions Analogies are made from pairs of words that have the same relationship. Complete the analogies using words from the box below.

snow	heat	coal	past	weather
sled	window	stove	storms	winter

1. HOT is to SUMMER as COLD is to _____.

2. RED is to RUBY as BLACK is to _____.

3. WASH is to SINK as COOK is to _____.

4. NOW is to THEN as PRESENT is to _____.

5. WOOD is to DESK as GLASS is to _____.

6. RICE is to ICE as SEAT is to _____.

Weather Watch

Directions Read the story. Use words from the box above to fill in the blanks.

People who live in northern countries have very cold

(7) _____. During the (8) _____ season,

they have a lot of (9) _____ and ice storms. Some storms

last for several days. These are called blizzards. When the

(10) _____ are over, children come outside to play. They

like to build snowpeople, ski, and (11) _____.

Subtracting Whole Numbers

REMEMBER

When you **subtract**, you take things out of groups. To show subtraction, you write −. This sign is called a **minus sign** or a **subtraction sign**. The answer is called the **difference**. Sometimes, you will need to **regroup**, or **borrow**, from another place value.

EXAMPLES

```
              9
          3 10 17  ◄────────        4 12   ◄──── regrouped numbers
   43       4 0 7                  $5.23
 − 21      −1 2 9                 −$1.80
 ────      ───────               ────────
   22       2 7 8                  $3.43
```

Find the Difference

Directions Find each difference. Regroup if you need to.

1. 78
    -41

2. 62
    -39

3. 80
    -51

4. 329
    $-\ 46$

5. 684
    -212

6. 915
    -389

7. $\$7.25$
    $-\$2.50$

8. $\$7.98$
    $-\$2.99$

9. $\$28.15$
    $-\ \$6.77$

Lines of Symmetry

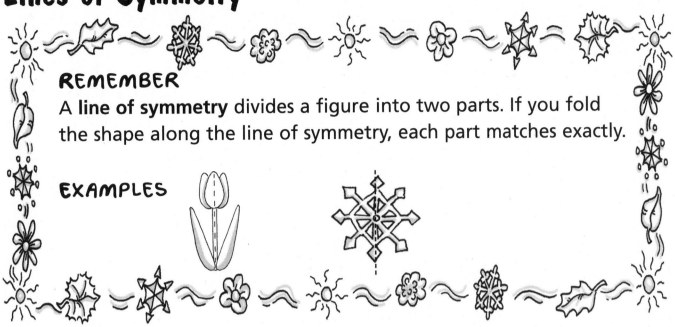

REMEMBER

A **line of symmetry** divides a figure into two parts. If you fold the shape along the line of symmetry, each part matches exactly.

EXAMPLES

Directions Circle **yes** if the dotted lines show symmetry. Circle **no** if they do not.

1.

yes or no

2.

yes or no

3.

yes or no

Directions Draw a line of symmetry for each figure.

4.

5.

6.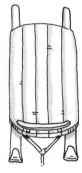

7. Draw a figure that does not have any lines of symmetry.

Weather Puzzle

Directions Read each definition. Use words from the box below to finish the puzzle.

sun	snow	thunder	wind
cloud	winter	tornado	rain

ACROSS

1. The star that gives light during the day
2. A storm with high winds
3. A cold season

DOWN

1. Frozen water
2. A loud noise made when clouds crash together
3. This is white or gray and floats in the sky
4. Air that moves across Earth

Sounds of the Season

Weather can come with many sounds. You may hear the crack of thunder when it rains, or your feet crunching in the fresh snow. List as many sound words you can about weather. Then write sentences describing those words.

_____

_____

_____

_____

_____

_____

_____

_____

_____

_____

_____

_____

_____

_____

_____

_____

The Boston Tea Party

On the night of December 16, 1773, the people of Boston had a tea party. They threw boxes of tea into the Boston Harbor. This event was called the Boston Tea Party.

Many people were upset with Britain. Britain's King George III put a tax on tea. People in other cities were upset, too. They did not want to pay the tax.

Many Americans did not want the British king to rule them. They wanted to make their own laws. This feeling was strongest in Boston.

The king gave them until midnight on December 16 to unload the tea. After midnight, the king's men would unload it themselves.

That night, about 50 men dressed in old rags. They kept their faces hidden. The men crept onto the ship. They dumped 300 boxes of tea into the ocean.

The Boston Tea Party

Directions Using what you have just read, answer the questions.

1. Why were the people of Boston upset?

   _____

   _____

   _____

2. Who was the king of Britain in 1773?

   _____

   _____

   _____

3. Why do you think that the people in Boston did not want to pay a tax?

   _____

   _____

   _____

4. Why do you think the men dressed in rags and hid their faces?

   _____

   _____

   _____

5. Why do you think this event is called The Boston Tea Party?

   _____

   _____

   _____

Make a Statement

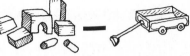

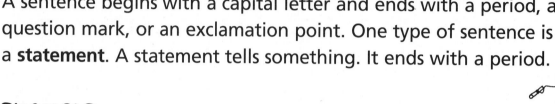

REMEMBER

A sentence is a group of words that tells a complete thought. A sentence begins with a capital letter and ends with a period, a question mark, or an exclamation point. One type of sentence is a **statement**. A statement tells something. It ends with a period.

EXAMPLE

I rode in a hot air balloon.

Directions Write each statement correctly.

1. a long time ago, a man sat in front of a fire

_____

2. he watched pieces of paper float up with the smoke

_____

3. the man got an idea

_____

4. he wanted to use heat to make a balloon fly

_____

5. the man made a plan for a hot air balloon

_____

6. his brother helped him

_____

7. a duck, a sheep, and a rooster were the first balloon passengers

_____

Rhyming Words

Directions Draw lines from the words on the left to the rhyming words on the right.

1. sent   belt

2. hear   dear

3. grow   rent

4. felt   stew

5. grew   show

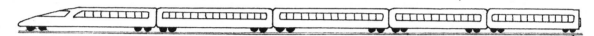

Directions Read the letter. Use words from the box below to fill in the blanks.

began	begin	feeling
built	heard	spend

Dear Editor:

When the town **(6)** _____ building the new train station,

I was all for it. I watched it being **(7)** _____. Then I

(8) _____ how much it was going to cost. Now my

(9) _____ is that the town planners are wrong. I think it

will cost the town too much money. I don't think we should

(10) _____ that much on trains! I think the people in the

town should **(11)** _____ to question if the station is really

worth it.

Sincerely,

Maya Taylor

Multiplying Whole Numbers

REMEMBER

To show **multiplication**, write ×. This sign is called a **multiplication sign**. The answer is the **product**. Sometimes, you will need to **regroup**, or **carry**, from another place value.

To multiply by one digit, multiply the top number one digit at a time. When a product is larger than 9, regroup, or carry, and continue multiplying.

- First, multiply the ones digit.
- Next, multiply the tens digit. Add any numbers you regrouped.
- Then, multiply the hundreds digit. Add any numbers you regrouped.

EXAMPLES

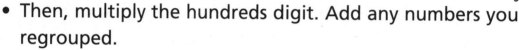
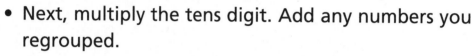

```
                    2  ←——   1 2  ←—regrouped numbers
      6            17        $3.26
    × 5           × 3        ×  4
    ————          ————       ————————
     30            51        $13.04
```

Let's Multiply!

Directions Find each product.

1.   9
      × 4

2.   8
      × 8

3.   7
      × 6

4.   32
      × 4

5.   54
      × 5

6.   60
      × 3

7. $2.71
     × 2

8. $4.36
     × 7

9. $5.18
     × 4

Follow the Pattern

Directions Use the pattern to answer the questions.

 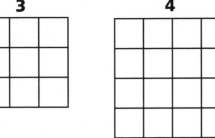

1   2   3   4   5

1. Look at the pattern. What figure would come next? Draw it in the space.

2. How do the patterns change from one figure to the next? Explain.

_____

_____

_____

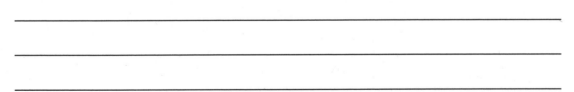

3. Without making a drawing, decide how many tiles will be in the tenth figure. Tell how you found this number.

_____

_____

_____

Then and Now

Directions Look at each picture. Draw a picture of what each item looks like today. Write a sentence telling how it has changed.

1.

_____

_____

2.

_____

_____

3.

_____

_____

The Best Invention

Directions Long ago, there were no televisions, radios, refrigerators, computers, or cars. Today, these things are part of our every day life. What do you think is the best invention? Write a paragraph telling why you think it is the best invention.

CHAPTER 9

The Fox and The Grapes

One summer day, Mr. Fox was walking in the woods. Soon his eyes grew very large. He saw big bunches of grapes growing on vines. The vines were growing near a tree. The branches reached high into the tree. These branches were heavy with grapes. How sweet the grapes looked! Mr. Fox liked to eat grapes. He was very excited!

Mr. Fox reached up to pick some grapes. He could not reach them. He tried and tried. Still, he could not reach the grapes. He jumped as high as he could. He still could not reach them.

Mr. Fox found a log. He placed the log beside the tree and climbed onto the log. The log rolled, Mr. Fox fell off, and he bumped his head on the tree trunk.

Mr. Fox grew tired. He looked at the grapes again. Then he turned away and started walking down the path. He said, "I really don't want those grapes. They probably don't taste good anyway!"

The Fox and the Grapes

Directions Using what you have just read, answer the questions.

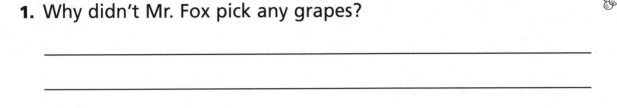

1. Why didn't Mr. Fox pick any grapes?

   _____

   _____

2. What happened when Mr. Fox climbed on the log?

   _____

   _____

3. Which sentence is a fact about the grapes? Circle it.

   These branches were heavy with grapes.

   How sweet the grapes looked!

4. How did Mr. Fox probably feel at the end of the story?

   _____

   _____

5. What does Mr. Fox need to learn?

   _____

   _____

Asking a Question

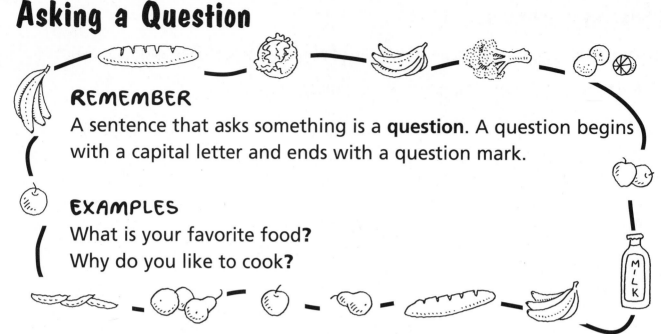

REMEMBER

A sentence that asks something is a **question**. A question begins with a capital letter and ends with a question mark.

EXAMPLES

What is your favorite food**?**

Why do you like to cook**?**

Directions Write each question correctly.

1. what are you cooking for dinner

_____

2. what time will we eat

_____

3. may I help you cook

_____

4. may I lick the spoon

_____

5. would you like milk to drink

_____

6. would you pass the bread, please

_____

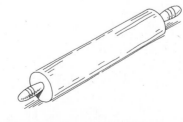

Word Groups

Directions Cross out the word that does not belong with the others.

1. stove   oven   drink

2. sugar   dishes   salt

3. knife   supper   breakfast

Baking Bread

Directions Read the story. Use words from the box below to fill in the blanks.

salt	bread	table
oven	sugar	baking

I spent the morning baking **(4)** _____ with my mother. We used flour, yeast, butter, and water to make it. We also used **(5)** _____ and **(6)** _____. We kneaded the dough on the kitchen **(7)** _____. Then we let it rise, kneaded it again, and put it in the **(8)** _____ to bake. When the bread was done, we let it cool on the counter. We had lots of fun **(9)** _____ bread.

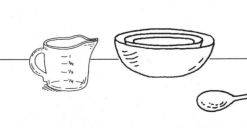

Dividing Whole Numbers

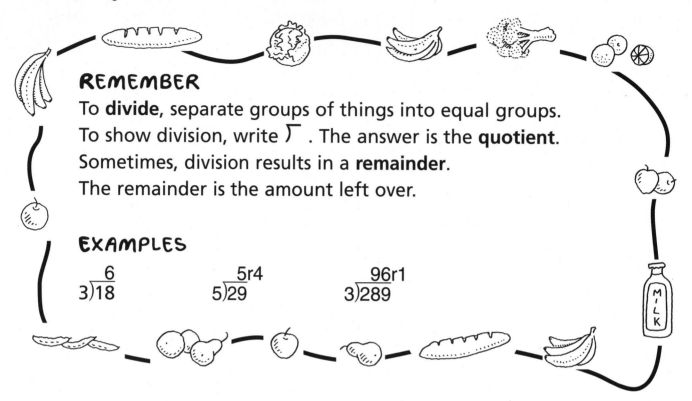

REMEMBER

To **divide**, separate groups of things into equal groups.
To show division, write ⌐ . The answer is the **quotient**.
Sometimes, division results in a **remainder**.
The remainder is the amount left over.

EXAMPLES

$$3\overline{)18}$$ = 6

$$5\overline{)29}$$ = 5r4

$$3\overline{)289}$$ = 96r1

Directions Find each quotient. Write the remainder if you need to.

1. $2\overline{)12}$ **2.** $7\overline{)28}$ **3.** $9\overline{)81}$

4. $5\overline{)39}$ **5.** $8\overline{)78}$ **6.** $4\overline{)73}$

7. $3\overline{)289}$ **8.** $5\overline{)507}$ **9.** $4\overline{)350}$

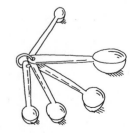

Cracker Stacker

Directions Joal works in a grocery store. He is setting up a display of crackers. He needs to put 24 boxes of crackers on the shelf. He wants to display the crackers in rows so that each row has the same number of crackers. Help Joal make the display by answering the questions.

1. If Joal displays the crackers in 3 rows, how many boxes of crackers can he put in each row?

   _____

2. If Joal wants the crackers in 8 rows, how many boxes of crackers can he put in each row?

   _____

3. Explain why a display that has 5 rows won't work for the cracker display.

   _____

   _____

4. Assume that Joal wants at least 3 rows, but no more than 8 rows. What four choices does Joal have for displaying the 24 boxes of crackers?

   _____

   _____

Food Scramble

Directions Read each clue. Unscramble the letters to find the answers. Write the words on the lines. Then use the numbered letters to solve the riddle.

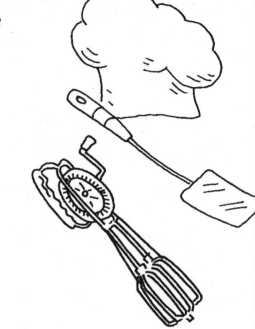

1. A purple or green fruit that grows on a vine

   prage   __ __ __ __ __
    1

2. A small, flat food that is usually sweet

   eokcoi   __ __ __ __ __ __
    2

3. A small, flat food that is usually salty

   erccark   __ __ __ __ __ __ __
    3

4. A baked food made with yeast that is used to make sandwiches

   drabe   __ __ __ __ __
    4

5. Something food is baked in

   nove   __ __ __ __
    5

Now solve the riddle:

What did the baby corn call his dad?

__ __ __   __ __ __ __
1 2 1   3 2 4 5

You Are What You Eat

Directions Have you ever heard the saying "You are what you eat?" Write sentences telling what you think this means. Then draw a picture of yourself showing what you ate yesterday. If you ate an apple, you might draw the apple as your head.

CHAPTER 10

The Long Drive

Brian pressed his forehead to the window. He had been sitting in the back seat for almost two hours now, but it was still morning. His family had packed up the car and left the house before the sun came up. They were going to Florida.

Brian had been doing anything to pass the time. He had already eaten the snack his mom packed for him. He had tried to read, but the words seemed to swim on the page. Brian could not sit still. He was too excited.

Grandpa Joe would be waiting for him. Grandpa Joe lived in Florida. Brian had not seen Grandpa Joe for almost a year. The two of them always had a good time together. Last time Brian was in Florida, Grandpa Joe took him to a basketball game. Grandpa Joe loves basketball.

"How much longer?" Brian asked. He hoped it was only a few more minutes.

"About two hours," his mom said. She was sitting in the front seat. She turned to face Brian. "Do you want to stop and stretch your legs?"

"NO!" Brian almost shouted. Stopping would make the trip longer. "Just keep going. No stopping," Brian said. He stared out the window anxiously.

The Long Drive

Directions Using what you have just read, answer the questions.

1. What is this selection mainly about?

   _____

   _____

   _____

2. What did Brian do to pass the time?

   _____

   _____

3. How long has it been since Brian has seen Grandpa Joe?

   _____

4. Why do you think that Brian did not want to stop to stretch his legs?

   _____

   _____

   _____

5. Is this selection real or make-believe? Tell how you know.

   _____

   _____

   _____

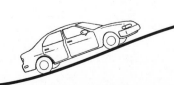

Exclamations!

REMEMBER

Some sentences show strong feelings such as surprise, fear, or excitement. A sentence that shows strong feeling is an **exclamation**. An exclamation begins with a capital letter and ends with an exclamation point.

EXAMPLES

Just look at that airplane**!**
What a beautiful old plane**!**

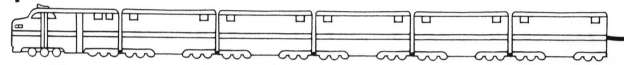

Show Your Excitement!

Directions Write each exclaiming sentence correctly.

1. look at that airplane fly

   _____

2. that pilot sits out in the wind

   _____

3. the pilot must get cold

   _____

4. that airplane is huge

   _____

5. this air show is so much fun

   _____

6. I could stay here all day

   _____

Making New Words

Directions Cross out one letter in each word to make a new word. Write the new word on the line.

1. near _____

2. close _____

3. around _____

4. where _____

5. below _____

6. left _____

Space Adventure

Directions Read the story. Use words from the box below to fill in the blanks.

on	close	Below
where	above	right

The crew didn't know **(7)** _____ the spaceship had

landed. They just knew it wasn't anywhere **(8)** _____ to

home. There were three glowing moons in the green sky up

(9) _____ . **(10)** _____ that was a purple

sea. The crew had landed **(11)** _____ a small island.

They felt safe until they saw a streak of lightning to the

(12) _____ of them. "We need to get out of here now!"

said the captain. "This place will not be safe when the storm comes."

Learning Large Numbers

REMEMBER

Large numbers have more than 4 digits.
In large numbers, commas separate digits into groups of three.
Notice each pattern of 100s, 10s, and 1s.

EXAMPLE

Millions			Thousands			Ones		
1	3	5,	7	8	0,	6	2	5
100s	10s	1s	100s	10s	1s	100s	10s	1s

To read a large number, read each group of digits separately.

135, 780, 625 is read 135 **million**, 780 **thousand**, 625.

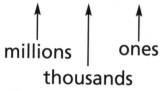

millions
thousands
ones

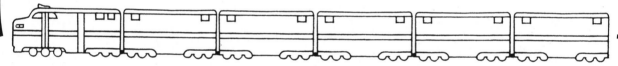

Directions Look at the number 235,794,016. Write the digit that is in each place.

1. thousands _____   2. ten thousands _____

3. millions _____   4. ten millions _____

5. hundred thousands _____   6. hundred millions _____

Directions Write each number using digits.

7. two hundred thirty-seven thousand _____

8. thirty-six million, nine hundred forty thousand _____

9. eight hundred million, two hundred thousand _____

10. one hundred thousand, five hundred nineteen _____

Using a Graph

Directions Use the graph to answer the questions.

1. What kind of graph is shown? Circle it.

   line graph   circle graph

2. Which is the largest age group?

   _____

3. Which is the smallest age group?

   _____

4. Which two groups make up about half of the United States' population?

   _____

   _____

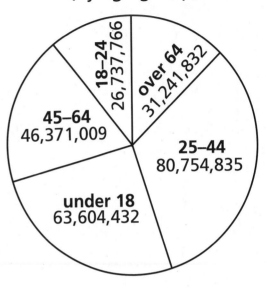

U.S. Population
(by age group)

18–24 26,737,766

over 64 31,241,832

45–64 46,371,009

25–44 80,754,835

under 18 63,604,432

5. Estimate how many more people are in the **under 18** age group than the **over 64** age group. Show your estimate.

   _____

6. Estimate the total number of people over the age of 44 in the United States. Show your estimate.

   _____

7. Estimate the total population of the United States. Show your estimate.

   _____

Road Codes

Directions Look at the codes. Each number stands for a letter. Write the letters on the blanks to answer the riddles.

A	B	C	D	E	F	G	H	I	J
1	2	3	4	5	6	7	8	9	10

K	L	M	N	O	P	Q	R	S	T
11	12	13	14	15	16	17	18	19	20

U	V	W	X	Y	Z
21	22	23	24	25	26

1. What kind of car is filled with water?

___ ___ ___ ___ ___ ___ ___ ___
1   3   1   18   16   15   15   12

2. What did the car say when it got new tires?

___ ___ ___ ___ ___   ___ ___ ___ ___ ___
9   6   5   5   12   23   8   5   5   12

___ ___ ___ ___.
7   15   15   4

3. How is a car like an elephant?

___ ___ ___ ___   ___ ___ ___ ___   ___ ___ ___ ___
20   8   5   25   2   15   20   8   8   1   22   5

___ ___ ___ ___ ___ ___.
20   18   21   14   11   19

4. When is a car not a car?

___ ___ ___ ___   ___ ___   ___ ___ ___ ___ ___
23   8   5   14   9   20   20   21   18   14   19

___ ___ ___ ___   ___
9   14   20   15   1

___ ___ ___ ___ ___ ___ ___ ___.
4   18   9   22   5   23   1   25

Take a Trip

Where would you like to go on a trip? How would you get there? What would you do once you got there? Draw a picture of a place you would like to visit. Then write sentences telling about your trip.

Jumping in Australia

Kangaroos are found only in the country of Australia. These animals have long legs, small heads, and strong tails. The biggest kangaroos are seven feet tall. The smallest kangaroos are about one foot tall.

Most kangaroos have red, brown, or gray fur. Female kangaroos are a darker color than male kangaroos. Female kangaroos are different in another way, too. They have a small pouch in front of their body. They use their pouch to carry baby kangaroos, called joeys.

Both male and female kangaroos can run fast. Their long legs help them jump too. Kangaroos have been known to leap thirty feet into the air! This makes the kangaroo one of the strongest animals in Australia.

Directions Fill in the bubble next to the correct answer.

1. Why is the kangaroo one of the strongest animals in Australia?

   Ⓐ They have a long tail.
   Ⓑ They have a small head.
   Ⓒ They have a pouch.
   Ⓓ They have long legs.

2. Which of the following sentences is true?

   Ⓐ Kangaroos have black fur.
   Ⓑ Baby kangaroos are called joeys.
   Ⓒ Most kangaroos are red.
   Ⓓ All kangaroos have pouches.

3. A kangaroo's tail is:

   Ⓐ long.
   Ⓑ flat.
   Ⓒ small.
   Ⓓ strong.

4. How are male and female kangaroos different?

   Ⓐ Females are a darker color.
   Ⓑ Males are a darker color.
   Ⓒ Females are larger.
   Ⓓ Males are faster.

Moving West

"It sure is hot," said Kerry as she wiped sweat from her brow.

"It sure is," said her father. He slowed the horses and guided them north.

Kerry looked worried. "Where are we going?" she asked.

"We are going to travel north for a while," he said, pointing to a forest.

"But I thought we were supposed to travel west," said Kerry.

Kerry's father chuckled. "You're right," he said. "But we cannot travel straight to the west. We have to change our direction. A large desert lies west of here. We couldn't make it through the desert without food and water."

"So we're going to get food and water in the forest!"

"Yes," her father answered.

Kerry smiled. She looked up at the blazing sun and then at her father. "He sure is a smart man," she thought. "I know he'll get us to the west safely."

Directions Fill in the bubble next to the correct answer.

5. Which happened first?

- Ⓐ Kerry asked her father why he changed direction.
- Ⓑ Kerry's father guided the horses north.
- Ⓒ Kerry's father pointed to the forest.
- Ⓓ Kerry wiped sweat from her brow.

6. Before the trip, Kerry probably thought they would:

- Ⓐ travel in many directions.
- Ⓑ travel straight west.
- Ⓒ travel through a desert.
- Ⓓ travel north.

7. What will Kerry and her father probably do next?

- Ⓐ They will get new horses.
- Ⓑ They will travel through a hot desert.
- Ⓒ They will change direction.
- Ⓓ They will hunt for food in the forest.

8. At the end of the story, Kerry is:

- Ⓐ worried.
- Ⓑ tired.
- Ⓒ relieved.
- Ⓓ afraid.

America's Game

Baseball is often called "America's Game." The game of baseball was not invented in America. The idea for baseball came from England. In the 1700s, children in England played a game called rounders.

In rounders, children took turns hitting a soft ball. When the batter hit the ball, he ran to a base. Other players would try to catch the ball. They would throw the ball at him. If the ball hit the batter, he was out.

In the next 100 years, people moving from England to America brought rounders with them. The game slowly changed. A hard ball was used and players had to be tagged with the ball to be called out. By the early 1900s, rounders had almost completely changed in America. Baseball had taken its place.

Directions Fill in the bubble next to the correct answer.

9. Which sentence is an opinion?

Ⓐ Baseball is often called "America's Game."

Ⓑ Baseball was not invented in America.

Ⓒ Baseball is more fun than rounders.

Ⓓ Rounders was played in England in the 1700s.

11. What is rounders?

Ⓐ A game like soccer

Ⓑ A game like baseball

Ⓒ A golf game

Ⓓ A football game

10. Rounders is different from baseball because:

Ⓐ A player is tagged out in rounders.

Ⓑ Players do not catch the ball in rounders.

Ⓒ Players run the bases in rounders, but not baseball.

Ⓓ A soft ball is used in rounders.

12. This passage is meant to:

Ⓐ tell how to play baseball.

Ⓑ explain the rules of rounders.

Ⓒ get more people to play baseball.

Ⓓ tell how baseball came to America.

City Helpers

Every city needs people who are helpers. Helpers make the city a better place to live. Some helpers also make the city safer too. Two city helpers are police officers and firefighters.

Police officers are needed in every city. They help people in danger. Police officers teach young people about bike safety. They also teach about how to stay safe going to and from school. Police officers are ready to help anytime.

Firefighters are also city helpers. They move very quickly when the fire alarm rings. They put out fires and save people from burning buildings. Firefighters also help people who are sick or have been in an accident. When people ask for their help, firefighters always answer the call.

Directions Fill in the bubble next to the correct answer.

13. What is this article mostly about?

- (A) firefighters
- (B) police officers
- (C) community helpers
- (D) teachers

14. Helpers make a city:

- (A) a more interesting place to live.
- (B) a dangerous place to live.
- (C) a more fun place to live.
- (D) a better place to live.

15. What happens when the fire alarm rings?

- (A) Firefighters move quickly.
- (B) Buildings begin to burn.
- (C) People answer the call.
- (D) Police officers save people.

16. Which of these is also a community helper?

- (A) a pilot
- (B) a mail carrier
- (C) a florist
- (D) a father

Stop! _____ out of 16 correct.

MATH CHECK-UP

Directions Read each question. Fill in the bubble next to the correct answer.

1. Which number means the same as forty-seven?

- (A) 74
- (C) 47
- (B) 470
- (D) 740

2. The chart below shows which number?

Thousands	Hundreds	Tens	Ones
2	0	1	8

- (A) 218
- (C) 201
- (B) 2,108
- (D) 2,018

3. The number **532** is the same as:

- (A) 5 hundred, 3 tens, 2 ones.
- (B) 5 tens, 3 hundreds, 2 ones.
- (C) 5 ones, 3 tens, 2 hundreds.
- (D) 5 hundreds, 3 ones, 2 tens.

4. Which number is greater than **125** but less than **150**?

- (A) 124
- (B) 152
- (C) 105
- (D) 149

5. Which number has an odd digit in both the hundreds and the ones place?

- (A) 211
- (C) 138
- (B) 367
- (D) 632

6. What number is more than **2,450** but less than **2,540**?

- (A) 4,250
- (C) 2,550
- (B) 4,520
- (D) 2,504

7. What is the value of 8 in the number **2,578**?

- (A) 8 ones
- (B) 8 tens
- (C) 8 hundreds
- (D) 8 thousands

8. Which number has a 3 in the thousands place?

- (A) 7,032
- (B) 3,072
- (C) 2,703
- (D) 7,302

MATH CHECK-UP

Directions Read each question. Fill in the bubble next to the correct answer.

9. $25 + 135 =$

- (A) 110
- (B) 150
- (C) 160
- (D) 385

10. $100 - 76 =$

- (A) 76
- (B) 24
- (C) 34
- (D) 176

11.
$$\begin{array}{r} 358 \\ + 212 \end{array}$$

- (A) 560
- (B) 650
- (C) 570
- (D) 146

12.
$$\begin{array}{r} 907 \\ - 50 \end{array}$$

- (A) 857
- (B) 957
- (C) 850
- (D) 840

13. $11 \times 6 =$

- (A) 6
- (B) 17
- (C) 60
- (D) 66

14. $9 \div 3 =$

- (A) 3
- (B) 6
- (C) 12
- (D) 27

15. $1\overline{)31}$

- (A) 30
- (B) 3
- (C) 31
- (D) 13

16.
$$\begin{array}{r} 375 \\ \times 4 \end{array}$$

- (A) 1,280
- (B) 1,300
- (C) 1,400
- (D) 1,500

MATH CHECK-UP

Directions Read each question. Fill in the bubble next to the correct answer.

The graph shows the number of cans Ms. Mark's class collected each week. Study the graph. Use the graph to answer questions 17 and 18.

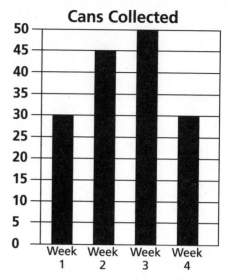

17. During which week were the most cans collected?

- Ⓐ Week 1
- Ⓑ Week 2
- Ⓒ Week 3
- Ⓓ Week 4

18. How many more cans were collected during Week 2 than Week 1?

- Ⓐ 15
- Ⓑ 10
- Ⓒ 5
- Ⓓ 0

19. Kim has a bag with 15 yellow crayons, 7 blue crayons, 5 green crayons, and 8 red crayons. If she takes one crayon out of the bag without looking, what color will she most likely pick?

- Ⓐ red
- Ⓑ green
- Ⓒ blue
- Ⓓ yellow

20. Roberto had 8 cookies. He and 3 friends each ate the same number of cookies. How many cookies did each person eat?

- Ⓐ 11
- Ⓑ 2
- Ⓒ 24
- Ⓓ 12

Stop! _____ out of 20 correct.

Congratulations!

_____

(name)

has completed *Summer Counts!*

Good job!

Have a good school year.

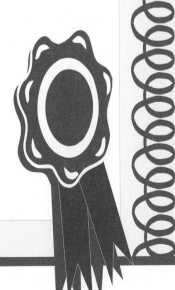

Answer Key

Page 5
1. It takes place in a shoemaker shop.
2. Customers bought the shoes.
3. They wanted to see who was making the shoes.
4. He wanted to thank the elves.
5. Make-believe. Elves are not real.

Page 6
1. classes
2. barn
3. animals
4. horse
5. puppies
6. chickens

Page 7
1. nurse
2. policeman
3. teacher
4. coach
5. fireman
6. classes
7. teacher
8. study
9. room
10. art

Page 8
1. four
2. eight
3. fifteen
4. eleven
5. fourteen
6. seventy
7. two hundred
8. three thousand
9. thirty
10. sixty
11. 3; 5; 8; 9
12. 6; 10; 14; 18; 20
13. 12; 18; 21; 30
14. 20; 40; 60; 80; 90
15. 10; 15; 30; 45; 50

Page 9
1. Ms. Reed's class
2. Mr. Starr's class
3. Mrs. James' class
4. About 15 pounds
5. About 105 pounds
6. 10 pounds

Page 10
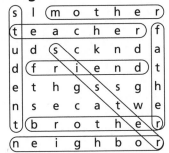

Page 13
1. It takes place in the woods.
2. The knight's armor made the noise.
3. He yelled, "Cut!"
4. Below them was a beautiful white horse.
5. Answers will vary. Accept reasonable responses.

Page 14
1. common: friend; proper: Rita
2. common: camp; proper: Sunday
3. common: camp; proper: Missouri
4. common: town; proper: Branson
5. common: month; proper: Camp Kano
6. common: parents; proper: Fourth of July

Page 15
1. lake
2. fishing
3. forest
4. beach
5. trail
6. family
7. camp
8. tent
9. boots
10. lake
11. fishing
12. vacation

Page 16
1. ones
2. hundreds
3. thousands
4. tens
5. thousands
6. 426
7. 852
8. 1,819
9. 2,604

Page 17
1. 34
2. 74
3. 347
4. 743
5. 734 or 374
6. 473, 743 or 347
7. You can make more odd numbers because there are two odd numbers and one even number.

Page 18
Across
3. backpack
5. trail
6. woods
7. tent

Down
1. hike
2. lake
3. boots
4. fire

Page 21
1. Possible answer: Cricket is lazy. He would rather play than work.
2. Ant works in the summer so he will have food for the winter.
3. Cricket looks for food on the ground.
4. Cricket learned not to wait until the last minute.
5. Make-believe. Animals do not talk.

Page 22
1. paddled; past
2. walk; present
3. follows; present
4. flipped; past
5. open; present
6. swim; present

Page 23
1. cat
2. swim
3. horse
4. pond
5. pond
6. ducks or geese
7. geese or ducks
8. bread

Page 24
1. 68¢
2. $1.00
3. $6.26
4. $3.57

Page 25
1. Yes, because the 7 cents must be made with 2 pennies.
2. 23 coins: 1 quarter and 22 pennies
3. 5 coins: 1 quarter, 2 dimes, and 2 pennies
4. 1 quarter, 1 dime, 2 nickels, and 2 pennies. Using a dime and two nickels instead of two dimes gives her one more coin.

Page 26
1. Put it on my bill.
2. She had tweets.
3. The pig squealed.
4. Cheap, cheap!

Page 29
1. 2, 3, 1, 4
2. Everyone cheered for Nick.
3. Possible answers: It means someone was about to get hurt, or was in danger.
4. Thick, black smoke was coming out of a window.
5. Answers will vary.

Page 30
1. is
2. was
3. is
4. are
5. were
6. was
7. are

Page 31
1. hello
2. reading
3. sent
4. word
5. news
6. newspaper
7. message
8. write
9. letter
10. reporter

Page 32
1. <
2. >
3. <
4. >
5. <
6. <
7. Top from left to right: 984; 926; 882; 820; 724.
Bottom from left to right: 943; 879; 835; 777; 709.

Page 33
1. 80°F
2. 84°F
3. Possible answer: Yes, because the temperature is rising 4° each hour. It will reach 92°F by 4:00 P.M.
4. Possible answer: It will reach 96°F at 5:00 P.M.

Page 34
1. note
2. letter
3. write
4. message
5. pencil
6. read
Riddle: a newspaper

Page 37
1. The article tells why children need to visit their own doctors.
2. Possible answers: Children need more calcium. They need to take less medicine.
3. Calcium makes bones strong.
4. It has calcium in it for bones.
5. Real. It gives facts about children and their doctors.

Page 38
1. She
2. him
3. It
4. She/He
5. them
6. We

Page 39
1. bone
2. nose
3. mouth
4. ear
5. eye
6. healthy
7. hands
8. hair
9. teeth
10. calcium
11. sick

Page 40
1. 50
2. 300
3. 4,000
4. 220
5. 700
6. 7,000
7. 30
8. 200
9. 2,800
10. 4,000

Page 41
1. 12 pencils
2. The blue pencil because there are more of them in the bag.
3. The red pencil because there are fewer of them in the bag.
4. Possible answer: Dr. Mendoza can take out 4 blue pencils or add 4 red pencils.
5. Dr. Mendoza should add 12 red pencils.

Page 42

Page 45
1. It is about a girl who is going to jump off a diving board.
2. It takes place at a swimming pool.
3. She was scared to dive off the board.
4. She takes a deep breath and holds her stomach.
5. Possible answer: Yes, she took a deep breath and held her nose. She probably dove in.

Page 46
1. most exciting
2. faster
3. warmer
4. most difficult
5. youngest
6. slower
7. more skillful

Page 47
1. dive
2. sleeping
3. swing
4. bat
5. team
6. won
7. win
8. pass/shoot/catch
9. pass/shoot/catch
10. pass/shoot/catch
11. run
12. sport

Page 48
1. 77
2. 118
3. 121
4. 189
5. 838
6. 1,000
7. $7.97
8. $13.97
9. $9.94

Page 49
1. 20
2. 4 miles; 22−18=4.
3. May: 20; June: 20
4. Possible answer: Juan will run about the same number of miles each month. This is reasonable because he has run that much for the past four months.
5. About 240 miles: Juan runs about 120 miles in 6 months, so he will probably run twice that in 12 months.

Page 50
1. high fly
2. fun run
3. wheel deal
4. base chase
5. great skate

Page 53
1. A twister is another name for a tornado.
2. There are strong winds, rain, and hail.
3. Possible answers: The storm knocked down power lines, people lost their homes.
4. It cost $450 million to clean up Fort Worth.
5. It cost so much because the tornado did a lot of damage.

Page 54
1. A
2. The
3. an
4. The
5. A
6. The
7. an
8. the

Page 55
1. winter
2. coal
3. stove
4. past
5. window
6. heat
7. weather
8. winter
9. snow
10. storms
11. sled

Page 56
1. 37
2. 23
3. 29
4. 283
5. 472
6. 526
7. $4.75
8. $4.99
9. $21.38

Page 57
1. yes
2. no
3. yes
4.
5.
6.
7. Figure should not show symmetry.

Page 58
Across
1. sun
2. tornado
3. winter
Down
1. snow
2. thunder
3. cloud
4. wind

Page 61
1. They were upset about the tax put on tea.
2. King George III
3. Possible answer: They did not want to give money to Britain.
4. They did not want to get caught.
5. Because they threw the tea into the ocean.

Page 62
1. A long time ago, a man sat in front of a fire.
2. He watched pieces of paper float up with the smoke.
3. The man got an idea.
4. He wanted to use heat to make a balloon fly.
5. The man made a plan for a hot air balloon.
6. His brother helped him.
7. A duck, a sheep, and a rooster were the first balloon passengers.

Page 63
1. rent
2. dear
3. show
4. belt
5. stew
6. began
7. built
8. heard
9. feeling
10. spend
11. begin

Page 64
1. 36
2. 64
3. 42
4. 128
5. 270
6. 180
7. $5.42
8. $30.52
9. $20.72

Page 65
1.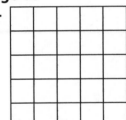
2. Possible answer: Each row and column increases by one.
3. 100 tiles will be in the square, because 10×10=100.

Page 69
1. Mr. Fox could not reach the grapes.
2. The log rolled and Mr. Fox fell off and bumped his head.
3. These branches were heavy with grapes.
4. Possible answer: Mr. Fox probably was sad because he could not get the grapes.
5. Possible answer: Mr. Fox needs to learn he should not give up so easily.

Page 70
1. What are you cooking for dinner?
2. What time will we eat?
3. May I help you cook?
4. May I lick the spoon?
5. Would you like milk to drink?
6. Would you pass the bread, please?

Page 71
1. drink
2. dishes
3. knife
4. bread
5. salt or sugar
6. sugar or salt
7. table
8. oven
9. baking

Page 72
1. 6
2. 4
3. 9
4. 7 r4
5. 9 r6
6. 18 r1
7. 96 r1
8. 101 r2
9. 87 r2

Page 73
1. 8 boxes of crackers
2. 3 boxes of crackers
3. The number of crackers on each shelf would not be even.
4. 3 rows of 8; 8 rows of 3; 4 rows of 6; 6 rows of 4.

Page 74
1. grape
2. cookie
3. cracker
4. bread
5. oven
Riddle: pop corn

Page 77
1. It is about a boy who is going to Florida on a car trip.
2. He ate his snack and read.
3. It has been almost a year since Brian has seen Grandpa Joe.
4. It would take longer to get there.
5. Real. A boy could really take a car trip to visit his grandfather.

Page 78
1. Look at that airplane fly!
2. That pilot sits out in the wind!
3. The pilot must get cold!
4. That airplane is huge!
5. This air show is so much fun!
6. I could stay here all day!

Page 79
1. ear
2. lose
3. round
4. here or were
5. blow
6. let
7. where
8. close
9. above
10. Below
11. on
12. right

Page 80
1. 4
2. 9
3. 5
4. 3
5. 7
6. 2
7. 237,000
8. 36,940,000
9. 800,200,000
10. 100,519

Page 81
1. circle graph
2. 25–44
3. 18–24
4. 25–44 and under 18
5. About 33 million; 64,000,000−31,000,000 =33,000,000.
6. About 77 million; 46,000,000+31,000,000 =77,000,000.
7. About 249 million.

Page 82
1. a car pool
2. I feel wheel good.
3. They both have trunks.
4. When it turns into a driveway.

Page 84
1. D
2. B
3. D
4. A

Page 85
5. D
6. B
7. D
8. C

Page 86
9. C
10. D
11. B
12. D

Page 87
13. C
14. D
15. A
16. B

Page 88
1. C
2. D
3. A
4. D
5. B
6. D
7. A
8. B

Page 89
9. C
10. B
11. C
12. A
13. D
14. A
15. C
16. D

Page 90
17. C
18. A
19. D
20. B